BLOCKCHAIN

THE ULTIMATE GUIDE TO UNDERSTANDING BLOCKCHAIN TECHNOLOGY

Charles Doyle

Copyright Notice.

©Charles Doyle

All rights reserved. No part of this publication may be reproduced, distributed or transmitted by any means or in any form, including but not limited to photocopying, recording, or other electronic or mechanical methods, without the prior written permission of the publisher, except in the case of brief quotations embodied in reviews and certain noncommercial uses acceptable to the copyright law.

Trademarked names appear in an editorial style without trademark symbols accompanying every occurrence of trademark names throughout the eBook. These names are used with no intention to infringe on the copyrights of respective owner trademarks. The information in this book is distributed on an "as is" basis, exclusively for educational purposes, without warranty. Neither the author nor the publisher shall have any liability to any person or entity with respect to any loss or damage caused or alleged to be caused directly or indirectly by the information contained in this book.

By reading this document, the reader agrees that Charles Doyle is under no circumstances responsible for any losses, direct or indirect, which are incurred as a result of the use of information contained within this document, including, but not limited to, —errors, omissions or inaccuracies.

Table of Contents

Introduction 1

Chapter 1: The History of Blockchain Technology 5

Chapter 2: The Mechanics of Blockchain 12

Chapter 3: Blockchain Applications 18

Chapter 4: Limitations of Blockchain 26

Chapter 5: Profiting from Blockchain 36

Chapter 6: Building a mining rig 42

Conclusion 49

Introduction

What is blockchain?

The introduction of cryptocurrencies, specifically Bitcoin, has brought the concept of blockchain technology into the mainstream. A blockchain is a continuously growing distributed database that protects against tampering and revision of data.

A blockchain is a group of transactions which are part of a common ledger that has been entered in a database and verified by multiple sources. Blockchain has more than one contributor and more than one transaction verification. In addition to this mind-boggling idea, Blockchain makes data secure and tamper proof, unlike physical or even electronic ledgers, which can be "cooked" by a scurrilous operator.

Blockchain's concept is similar to building a house of Legos. To build the house, you must begin with the foundation. Once the foundation is laid, the builder confirms everything is in place and is capable of supporting the next layer. The engineer looks for flaws or mistakes in the layout, the construction and even the strength of the foundation. When the engineer is sure the foundation is trustworthy, he calls in the building inspector to confirm that everything is in fact correct.

The similarity between Blockchain and Legos is demonstrated by ensuring that each piece of data added to the blockchain is individually crafted and tested, like bricks of Legos. No single person inputs into the database and verifies their own information. The data contributor is identified by the Internet Protocol (IP) address, and the verifier is identified by the IP address. The information is anonymous and coded so that the data entry operator does not know what they are processing. They don't even have complete information about the transaction, so they aren't able to abscond with the personal details of a client and commit identity theft.

BLOCKCHAIN

In addition to the transaction being secure, the transaction is permanent, and cannot be altered by either an inside or an outside source. Each block of data is linked to the previous block of data by a timestamp which contains the date, time, and origin.

The blockchain is a public ledger in which anyone can see the transactions and anyone can see the transactions and anyone can add to the ledger. Many parties share the same foundation and are making multiple entries at the same time, adding them to the bricks of the previous layer or foundation.

The Bitcoin project started doing rounds on the internet in the early 2010. One could then be interested in this project since implementing a digital currency brings about numerous challenges that have plagued traditional currencies for years.

There are issues of information sharing, a lack of payment processors, and the numerous problems surrounding counterfeiting this currency. Put simply, while Bitcoin might seem simple, the implementation had to be nothing short of genius for the system to truly work.

This is the first foray into Blockchain technology, the underpinnings of what powers and supports Bitcoin and other cryptocurrencies. In trying to develop a product that mimics gold, a group of programmers created an open source project that solves problems they never even realized they were working on.

Cryptocurrencies might be interesting, and there is even profit to be had if you do your own mining, but this pales in comparison to the numerous ways in which Blockchain technology will change the way information is shared across the world.

There is no doubt that Blockchain technology will become a central part of our lives over the next ten to fifteen years. Soon you will start to see large financial institutions use the technology to revamp the way that they create and are held to contracts. Blockchain will be massively disruptive for payment processors and credit cards. It is a very real possibility that not too long in the future a government will adopt Blockchain as the underpinning of their official currency.

Motivation for Blockchain Technology

Traditional databases are maintained by a single organization, and that organization has complete control of the database, including the ability to tamper with the stored data, to censor otherwise valid changes to the data, or to add data fraudulently. For most use cases, this is not a problem since the organization which maintains the database does so for its own benefit, and therefore has no motive to falsify the database's contents; however, there are other use cases, such as a financial network, where the data being stored is too sensitive and the motive to manipulate it is too enticing to allow any single organization to have total control over the database. Even if it could be guaranteed that the responsible organization would never enact a fraudulent change to the database (an assumption which, for many people, is already too much to ask), there is still the possibility that a hacker could break in and manipulate the database to their own ends.

The most obvious way to ensure that no single entity can manipulate the database is to make the database public, and allow anyone to store a redundant copy of the database. In this way, everyone can be assured that their copy of the database is intact, simply by comparing it with everyone else's. This is sufficient as long as the database is static; however, if changes must be made to the database after it has been distributed, a problem of consensus arises: which of the entities keeping a copy of the database decides which changes are allowed and what order those changes occurred in? If any of the entities can make changes at any time, the redundant copies of the database will quickly get out of sync, and there will be no consensus as to which copy is correct. If all of the entities agree on a certain one who makes changes first, and the others all copy from it, then that one has the power to censor changes it doesn't like. Furthermore, if that one entity disappears, the database is stuck until all of the others can organize to choose a replacement. All of the entities may agree to take turns making changes and all the others copy changes from the one whose turn it is, but this opens the question of who decides who gets a turn when.

Blockchain has truly moved out of the realm of interesting and into the spotlight as a practical, powerful and necessary tool to facilitate information sharing in the twenty-first century.

Over the following chapters, you will learn about the fundamentals of Blockchain technology. Explanation will be given in laymen language how the technology works, and why it is so revolutionary. You will discover the many uses of Blockchain technology from the fist cryptocurrencies to the groundbreaking projects being powered by Blockchain today.

You will see how the future will be affected by Blockchain technology, and you will even learn how you can make profit through Blockchain technology and mining today.

Continue reading and soon you will have a complete understanding of Blockchain technology. From its history to the mechanics of its function, to even understanding how to profit from the system; this is a one-stop guide for everything you need to know about Blockchain.

Chapter 1: The History of Blockchain Technology

Many of the technologies we now take for granted were quiet revolutions in their time. Just think about how much smartphones have changed the way we live and work. It used to be that when people were out of the office, they were gone, because a telephone was tied to a place, not to a person. Now we have global nomads building new businesses straight from their phones. And to think: Smartphones have been around for merely a decade.

We're now in the midst of another quiet revolution: blockchain, a distributed database that maintains a continuously growing list of ordered records, called "blocks." This innovation landscape represents just 10 years of work by an elite group of computer scientists, cryptographers, and mathematicians.

These changes and others represent a pervasive lowering of transaction costs. When transaction costs drop past invisible thresholds, there will be sudden, dramatic, hard-to-predict aggregations and disaggregation of existing business models. For example, auctions used to be narrow and local, rather than universal and global, as they are now on sites like eBay. As the costs of reaching people dropped, there was a sudden change in the system. The blockchain is reasonably expected to trigger as many of these cascades as e-commerce has done since it was invented, in the late 1990s.

Predicting what direction it will all take is hard. Did anybody see social media coming? Who would have predicted that clicking on our friends' faces would replace time spent in front of the TV? Predictors usually overestimate how fast things will happen and underestimate the long-term impacts. But the sense of scale inside the blockchain industry is that the changes coming will be "as large as the original invention of the internet," and this may not be overstated. What we can predict is that as blockchain matures and more people catch on to this new mode of collaboration, it will extend into everything from supply chains to probably fair internet dating (eliminating the possibility of fake profiles and other underhanded techniques). And given how far blockchain come in 10 years, perhaps the future could indeed arrive sooner than any of us think. This is all where it started:

BLOCKCHAIN

Digital gold

It is impossible to discuss Blockchain without mentioning Bitcoin. Bitcoin is by and far the most recognized use of Blockchain technology. It is the foundation for the how and why to digital currencies. Interestingly, the history of Blockchain starts with trying to solve an information problem; that is, information needs to be equally distributed across participants in the network.

How this problem relates to the building of a digital currency all starts with the idea that the monetary system is controlled by the few, to the detriment of the many.

Consider for a moment a single piece of currency on your persons. Whether that is a coin or a bill, it does not matter. It does not need US tender either; what is important is that a government issued it. Now consider the question: "what gives this piece of currency value?"

You know you can exchange it for goods and services, but the currency in and of itself, it represents something larger. It shows the faith based monetary system that the world has adopted. Essentially, what gives the dollar in your pocket power is the idea that others will be able to use that dollar for their own purposes. The person that accepts the tender has faith in the system as a whole and believes that the currency will not change value, nor become invalid before the time they use it.

This was not always the case with currency, and it actually brought a relatively recent change in the monetary policy. The currency was originally backed by something material, gold. The creators of Blockchain believed in gold backed currency to the degree that this is what Blockchain technology is meant to emulate- a gold backed system in all its strengths and weaknesses.

There are a group of advocates for a gold backed currency system. Their belief is that by providing the power to manipulate currency in the hands of the government, you open up the platform to human error. Their main concern is one of inflation; or rather the declining purchasing power of a dollar over time.

This is a matter of fact in any monetary system; as the economy grows and productivity increases, the relative value of pieces of currency will change appropriately. Without an additional injection of currency into the system, ideally, each piece of currency will change appropriately. Without an additional injection of currency into the system, ideally, each piece of currency would grow in value. This makes sense, as the currency becomes more valuable because the total supply is limited.

It is in the avoidance of human error, and the desire to create a gold-like currency that Bitcoin and Blockchain Technology was created.

Think for a moment about why gold has value, to begin with. It is not the cost to manufacture the gold, nor is it the cost to find and dig up the gold. And yet gold continues to climb in value. This is attributed solely to the scarcity of gold.

More gold cannot be created, and the likelihood of a large gold deposit being found is unlikely. Gold has value merely because it is in limited supply, and so for a currency system without human intervention, it is nearly perfect.

One person cannot find more gold, meaning that the power of each ounce of gold maintains relatively stable over time. Granted large-scale financial crises are much harder to work through with a gold standard, but the basic nature of its rarity, limited supply, and impossibility to counterfeit lend credence to the idea that a gold backed system is the best possible monetary system.

Right now the idea of gold and monetary systems might seem loosely connected to Blockchain technology as a whole, but it is in addressing the problem of replicating gold-like system that Blockchain technology was created. A group of programmers online needed to create a system that mimicked gold. Some of their solutions were novel, and others were completely brilliant.

BLOCKCHAIN

It starts with the idea of how gold is introduced into a system. It is not merely all dumped at one time. The history of civilization has shown that gold was found very quickly at first, but as more and more deposits were exhausted, the amount of gold flowing into the system became more limited. To mimic this system you would need to create a digital substitute.

The greatest problem then would be how to control the creation of this digital currency, mimicking the quick rise of gold in the beginning, to the slow taper of gold that is found later. Additionally, how could one avoid the counterfeiting of this digital currency?

Most importantly, how can transactions be counted and known to all parties, meaning if you give a Bitcoin, how do you know you have received it, and what is stopping you from giving that same Bitcoin to another person. Without any central financial institutions, these problems are quite difficult to solve.

The answer lies in a decentralized system where anonymous parties force cooperation through a Blockchain system. When a Bitcoin is transferred, every single person on the Bitcoin network knows it. When a coin is created, everyone in the network knows this too. With this system, the need for an overarching financial system is reduced. There is no reason why an individual can't cooperate if everyone always knows what is in each other's pockets- this is the genius of a Blockchain system.

It allows for information to be recorded by multiple parties, and the technology allows for all of this information to be recorded exactly the same. It cannot be modified or changed without every single person in the network is made aware of these changes. In addition, changes are made by a majority rule, meaning that if one person claims to have created a Bitcoin, but others in the network disagree, the majority will overrule and record that the Bitcoin, in fact, did not get created, and was not attributed to this person.

The genius of this system then is that it forces cooperation between the many and eliminates the possibility of a single party to cheat the system. With no one in control, truly the entire network and community have power over the currency.

With a working decentralized system, there is one last issue to solve, the problem of how much currency should be created. The answer to this last question is perhaps the most genius, as it incorporates keeping track of transactions while creating currency.

The verification that transfers of Bitcoin have taken place is calculated by 'miners', or those that are using computer hardware to verify transactions, and approximately every ten minutes a number of transactions are verified. The person that is able to create this verification is rewarded with some Bitcoin. The gold model represents the reverse exponential growth of Bitcoin.

If you imagine a standard parabola using a basic exponential algebraic graph, flip it on its head and you will see the roadmap for Bitcoin generation. The most Bitcoin will be generated in the beginning, resembling finding large gold deposits. Over time the total amount of Bitcoin that floods the market will become less and less. There is a cap to the total number of Bitcoins that can be created- this number is twenty one million.

After this number has been met, no more Bitcoins will ever be created. Once again, this idea comes from the basis of the gold standard. With a limited supply, just like any commodity, the standard Bitcoin would grow in value. A single agency or government body would not determine inflation; the free market would determine how much a Bitcoin is worth.

In later chapters, you will learn more about the Blockchain system and how it facilitates information sharing that it is not controlled by a single party. For right now, you should appreciate where this technology came from. Just like many others great inventions, the creators of Blockchain were not trying to revolutionize information sharing.

They were striving for something much more basic, but in solving a complex problem they have left us with a technology as revolutionary as the early internet protocols of the late 1970s. The blockchain is just as revolutionary a technology today- it will be the protocol for exchanging information in the future.

A technology of the internet era

One of the most fascinating aspects of Blockchain technology is that it truly is a creation of the internet. It is not precisely known who created Blockchain, but rather the only information we have is an online alias, Satoshi Nakamoto. There has been much debate around who this central character is. There is good evidence to support that Nakamoto is, in fact, a professor at a university in Australia, and yet still we are not certain of the true identity of this elusive figure.

This differs greatly from the creation of the early internet, where TCP/IP protocols were designed by the military and private enterprise. Nakamoto was the first to propose the idea of the Blockchain, but the programming was a completely open source project- the idea nevertheless will forever be attributed to Nakamoto.

It is in the open nature of the Bitcoin project that we have the development of Blockchain. One of the fundamental ideas behind the technology was that the people creating it did not have true trust in each other. If there was any money to be made from this project, each was worried that one person would steal money from the other members in the party.

From this background, we have a technology with the basic premise that other people cannot be trusted. This may seem overly negative, but as a rule in economics, it is precise. Exchanging tender today works because we have faith in the currency that is being handed to us- we don't need to worry about the reputation of who is handing us money because we trust the money itself. This was the problem behind Bitcoin, and Blockchain is the answer of restoring faith to the network.

While in theory, the creators of Bitcoin made a remarkable technology out of the desire to have an ideal currency, in all likelihood at least some of them got very rich. Keep in mind how Bitcoins are created, through computation increases, meaning that early on it was relatively easy to create Bitcoins.

Additionally, the network was so small that your odds of mining coins were far greater than they are today. As far as we can tell the Bitcoin system was running for some time before the network launched to the public. This means that it is likely Nakamoto has several thousand Bitcoins to him or herself.

The evolution of Blockchain to additional applications is also of the internet era. For years after the launch of Bitcoin, the fascination came from the currency itself, and the libertarian paradise that a gold standard could bring.

It wasn't until the source code was well documented that the brilliance of Blockchain was realized. As a solution for currency, Blockchain is essential. As a solution to sharing information across disparate parties, Blockchain is truly revolutionary. The applications were slow to start, but since 2015 there has been a great increase in the amount of interest of Blockchain as a protocol for all types of products. In the coming years, you will see more and more uses of Blockchain, and it is likely to be a huge factor in your life moving forward. Whether you notice the underpinning technology or not, Blockchain will be there.

BLOCKCHAIN

Chapter 2: The Mechanics of Blockchain

A permanent record

To best understand how blockchain technology works, we can start with a very simple example. Suppose for a moment that you and two friends came across three magic sheets of paper.

What was the magic about these pieces of paper was that anything written on one sheet would immediately appear on the other two sheets of paper? If what was written was erased, it would reappear on the same sheet of paper. The only way to erase information was if two people erased the same markings on two sheets of paper at the same time. This, in essence in the magic behind blockchain technology, but it does not explain its true use.

There are many practical applications for how these magic sheets of paper could be used, but for you and your friends, you decide on something very simple; to keep track of money owed between the three of you.

Using magic sheets of paper for the purpose of keeping track of debts may seem pretty, but this is a true problem on a larger scale, and even with friends it can cause tension and reason for argument.

With your two friends, whenever one person lends money to another person, they write this transfer down on the magic sheet of paper, causing the instant transmission of this information to other two sheets. What is important to realize about this premise is that a third party is able to see a transfer between two individuals, even if they did not participate.

This means that the third party is made aware of the debt. In a complex set of debts, where one person lends to another, and then another person borrows from that person, we create a system where paying back one another is made easier. For example, if you borrow from person A, and then person B borrows from you, person B could pay back person A instead of paying you back. This is a simple process but think about all of the points of confusion that could happen.

By bypassing the initial borrower, that person could claim that they never got paid back. The magic sheets of paper would show that this isn't true, and as the information is shared every member in the party, everyone knows what money is owed and to whom.

Also important in this situation is the ability for the majority to veto a transaction if it is disputed. For example, if one friend claims to be owed money and writes down that another borrowed from them, the other two can cross out this transaction.

Assume for a moment that two friends wouldn't try and cheat another and you can see that disputing information this way works to allow equal information for all parties. No one can cheat another, and all transactions can be verified by looking at the magic sheet of paper. Also, note that transactions that are paid back are not erased in this system; they are merely checked off.

The only markings that are ever erased are the ones that are not truthful, so all transactions back and forth are recorded and the only thing erased is information that is disputed by two of the three members of the group.

Using magic sheets of paper for this purpose is a microcosm of how blockchain technology works. Imagine now that instead of you and two friends, one hundred people have these same magic sheets of paper. These one hundred people can lend money to each other, and write it down on this sheet of paper the same way. Transactions that are disputed can only be erased by a majority of the holders of the magic sheets of paper. It ensures trust in this larger group, even among strangers. Everyone benefits because it is in everyone's best interest to stamp out fraud.

Incorrect transactions hurt the entire group, but the motivation to eliminate fraud is not based on a desire to help the group, it is based on a desire to benefit each individual member of the group. By working within a system that understands the inherit bias that all people have to value themselves above others, we have a system that is accountable to all members of the group without issuing more power to one person within that group.

The distributed ledger

The magic sheets of paper in the previous example hold the basic idea behind the distributed ledger that is central to the blockchain. Instead of magic sheets of paper, each transaction or exchange of information is recorded in a long chain of data blocks. These form the 'blocks' of a blockchain.

Relative to the previous example, you can think of each transaction added to the magic sheets of paper as a single block. In reality, blocks contain many transactions, are identified by their time stamp. For example, with bitcoin, each block represents roughly ten minutes of transactions. The blocks link to each other to form a whole picture that has ever taken place.

Each person within the blockchain network has access to all of the previous blocks of data. This chain serves as a ledger of all of the transactions and is held by each member of the network.

In this system, it is impossible for one person to cheat the group. If they modify their ledger to create currency, or write that someone owes them money, the group as a whole can see that this transaction is invalid. There needs to be an agreement between all members of the network for any changes to take place. It is a simple idea but one with great ramifications in terms of cooperation.

Take for instance how payment transactions are recorded today. Most transactions must go through a third party payment processor when goods are purchased online. Whether that processor is PayPal, MasterCard or Visa, there is a central power that holds a ledger for all transactions. This is not something that you are privy to, and your ability to dispute transactions is rather limited. Most importantly, you cannot look at a separate copy of the ledger to identify errors that may have occurred in the past.

BLOCKCHAIN

There is a good reason why distributed ledger cannot work in today's system. The answer lies in privacy and the identifying information that is present in all transactions. The information collected by a payment processor is the name of the purchaser, plus the credit card or PayPal information that can be used over and over again for purchases.

Handing this information out is dangerous, and it makes sense that a central party would hold te information. This is one of the central problems that blockchain tries to solve. There needs to be a sense of anonymity behind transactions. After all, the ledger will be in the hands of everyone within the network.

The network simply cannot afford to have identifying personal information be part of the record. The solution to this problem lies in emulating the exchange of physical currency.

Bitcoin and transfers

Anonymity is a central aspect to the blockchain, but more importantly to cryptocurrencies in general. In addition to trying to mimic gold, these currencies are trying to mimic cash as well. Without a third party payment processor that handles identifying information, it is possible for a transfer to take place between two parties without anyone else knowing a specific person made a transaction.

In the blockchain, the information recorded is the wallet numbers and the exchange of coins between these wallets. The persons behind these numbers are private, so while transactions are all recorded, the actual people behind transactions remains anonymous.

The system works by issuing public wallet identification number along with a private key that unlocks a digital wallet. For example, suppose that you wanted to purchase a good online using Bitcoin or another cryptocurrency.

The standard method for this transaction would be to provide to the merchant your public wallet number, along with the merchant providing you with their wallet number. By using your private key with your public number, you are authorized to transfer coins from your wallet to theirs. I the block of data that gets added to the greater chain, the information recorded are the public wallet numbers along with the transfer of coins from one wallet to another.

This system has its advantages in the anonymity it provides, but also brings about a problem common to uses cash currency; it is relatively easy for you to lose your account. Just like losing your wallet is only thing barring someone from using your cryptocurrency.

In addition, if you happen to lose this key you will no longer be able to assess your wallet. This aspect brings a sense of physicality to the bitcoin network, emulating a wallet in nearly every way. It has all of the advantages and disadvantages of physical currency.

As a new user, you can get started with Bitcoin without understanding the technical details. Once you have installed a Bitcoin wallet on your computer or mobile phone, it will generate your first Bitcoin address and you can create more whenever you need one. You can disclose your addresses to your friends so that they can pay you or vice versa. In fact, this is pretty similar to how email works, except that Bitcoin addresses should only be used once. The explanation of how the blockchain technology works in a nutshell:

Balances - block chain

The block chain is a shared public ledger on which the entire Bitcoin network relies. All confirmed transactions are included in the block chain. This way, Bitcoin wallets can calculate their spendable balance and new transactions can be verified to be spending bitcoins that are actually owned by the spender. The integrity and the chronological order of the block chain are enforced with cryptography.

Transactions - private keys

BLOCKCHAIN

A transaction is a transfer of value between Bitcoin wallets that gets included in the block chain. Bitcoin wallets keep a secret piece of data called a private key or seed, which is used to sign transactions, providing a mathematical proof that they have come from the owner of the wallet. The signature also prevents the transaction from being altered by anybody once it has been issued. All transactions are broadcast between users and usually begin to be confirmed by the network in the following 10 minutes, through a process called mining.

Processing - Mining

Mining is a distributed consensus system that is used to confirm waiting transactions by including them in the block chain. It enforces a chronological order in the block chain, protects the neutrality of the network, and allows different computers to agree on the state of the system. To be confirmed, transactions must be packed in a block that fits very strict cryptographic rules that will be verified by the network. These rules prevent previous blocks from being modified because doing so would invalidate all following blocks. Mining also creates the equivalent of a competitive lottery that prevents any individual from easily adding new blocks consecutively in the block chain. This way, no individuals can control what is included in the block chain or replace parts of the blockchain to roll back their own spends.

Chapter 3: Blockchain Applications

You should now understand the premise to blockchain and why it is a useful technology. It forces participation among anonymous parties and distributes power over this information to the entire blockchain network. This technology was created as a solution to a monetary problem, but with the solution, we now have a technology that can address many other problems.

Since 2015 there has been an explosion in the practical uses of blockchain technology, going as far that you can create your own blockchain with relative ease. As time goes on the uses for a distributed ledger will grow larger, and below are simply the uses for blockchain that have gained the most transaction over the last few years.

- ❖ **Financial Services**

Traditional systems tend to be cumbersome, error-prone and maddeningly slow. Intermediaries are often needed to mediate the process and resolve conflicts. Naturally, this costs stress, time, and money. In contrast, users find the blockchain cheaper, more transparent, and more effective. A few people wonder that a growing number of financial services are using this system to introduce innovations, such as smart bonds and smart contracts. The former automatically pays bondholders their coupons once certain preprogrammed terms are met. The latter are digital contracts that self-execute and self-maintain, again when terms are met.

Examples of blockchain financial services

- Asset Management: Trade Processing and Settlement

Traditional trade processes within asset management (where parties trade and manage assets) can be expensive and risky, particularly when it comes to cross-border transactions. Each party in the process, such as broker, custodian, or the settlement manager, keeps their own records which create significant inefficiencies and room for error. The blockchain ledger reduces error by encrypting the records. At the same time, the ledger simplifies the process, while canceling the need for intermediaries.

- Insurance: Claims processing

Claims processing can be a frustrating and thankless procedure. Insurance processors have to wade through fraudulent claims, fragmented data sources, or abandoned policies for users to state a few – and process these forms manually. Room for error is huge. The blockchain provides a perfect system for risk-free management and transparency. Its encryption properties allow insurers to capture the ownership of assets to be insured.

- Payments and Money Transfers

Perhaps the most well-known Blockchain application is being able to send and receive payments. Since blockchain technology has its beginnings in cryptocurrency, this makes sense. But, how exactly is this beneficial for small business owners?

By using blockchain technology, you're able to transfer funds directly and securely to anyone you want in the world almost instantly and at ultra-low fees. That's because there aren't any intermediaries slowing down the transfer of funds between several banks and charging outrageous transaction fees. This practice is especially useful if you have remote employees or are involved in the global marketplace.

❖ Smart contracts

These are legally binding programmable digitized contracts entered on the blockchain. What developers do is to implement legal contracts as variables and statements that can release of funds using the bitcoin network as a '3rd party executor', rather than trusting a single central authority.

For example, if two people want to exchange $100 at a specific time in future when a set of preconditions are met, the conditions, payout, and parties' details would be programmed into a smart contract. Once the defined conditions are met, funds would be released and sent to the appropriate party as per terms.

By giving computers control over contracts, we can make business more efficient and make the legal system more equitable.

Examples of Blockchain Smart Contracts

- Blockchain Healthcare

Personal health records could be encoded and stored on the blockchain with a private key which would grant access only to specific individuals. The same strategy could be used to ensure that research is conducted via HIPAA laws (in a secure and confidential way). Receipts of surgeries could be stored on a blockchain and automatically sent to insurance providers as proof-of-delivery. The ledger, too, could be used for general health care management, such as supervising drugs, regulation compliance, testing results, and managing healthcare supplies.

- Blockchain music

Key problems in the music industry include ownership rights, royalty distribution, and transparency. The digital music industry focuses on monetizing productions, while ownership rights are often overlooked. The blockchain and smart contracts technology can circuit this problem by creating a comprehensive and accurate decentralized database of music rights. At the same time, the ledger and provide transparent transmission of artist royalties and real time distributions to all involved with the labels. Players would be paid with digital currency according to the specified terms of the contract.

- Blockchain Government

In the 2016 election, Democrats and Republicans questioned the security of the voting system. The Green Party called for a recount in Wisconsin, Pennsylvania, and Michigan. Computer scientists say hackers can rig the electronic system to manipulate votes. The ledger would prevent this since votes become encrypted. Private individuals can confirm that their votes were counted and confirm who they voted for. The system saves money, by the way, for the government, too.

❖ Smart Property

A tangible or intangible property, such as cars, houses, or cookers, on the one hand, or patents, property titles, or company shares, on the other, can have smart technology embedded in them. Such registration can be stored on the ledger along with contractual details of others who are allowed ownership in this property. Smart keys could be used to facilitate access to the permitted party. The ledger stores and allows the exchange of these smart keys once the contract is verified.

The decentralized ledger also becomes a system for recording and managing property rights as well as enabling the smart contracts to be duplicated if records or the smart key is lost. Making property smart decreases your risks of running into fraud, mediation fees, and questionable business situations. At the same time, it increases trust and efficiency.

Examples of Blockchain Smart Property.

- Unconventional money lenders/ hard money lending

Smart contracts can revolutionize the traditional lending system. For instance, unconventional money lenders (e.g. hard money lenders) service borrowers who have poor credit with needed loans – while charging two to ten percent of the loan amount and claiming their property as collateral. Too many borrowers fall into bankruptcy and lose homes. The blockchain can undercut this by allowing a stranger to loan you money and taking your smart property as collateral. No need to show the lender credit or work history. No need to manually process the numerous documents. The property's encoded on the blockchain for all to see.

- Your car/ smartphone

BLOCKCHAIN

Primitive forms of smart property exist. Your car key, for instance, may be outfitted with an immobilizer, where the car can only be activated once you tap the right protocol on the key. Your smartphone too will only function once you type in the right PIN code. Both work on cryptography to protect your ownership.

The problem with primitive forms of smart property is that the key is usually held in a physical container, such as the car key or SIM card, and can't be easily transferred or copied. The blockchain ledger solves this problem by allowing blockchain miners to replace and replicate a lost protocol.

- Blockchain Internet-of-Things (IoT)

Any material object is a 'thing.' It becomes an internet of things (IoT) when it has an on/ off switch that connects it to the internet and to each other. By being connected to a computer network, the object, such as a car, become more than just an object. It is now people-people, people-things, and things-things.

How does the IoT affect you? Your printer can automatically order cartridges from Amazon when it runs low. Your alarm clock will change your time for brewing coffee, while your oven will produce an immaculately timed turkey for Thanksgiving. These are just some examples. On a larger scale, cities and governments can use IoT to develop cleaner environments, more efficient energy use and so-called 'smart cities,' to improve how we live and work.

❖ Shared Databases

You may have heard news organizations claiming that 2017 marks the beginning of the end of "facts". It is impossible to verify anything as true because there is so much information on the internet that truth claims are often considered false, while false claims are often remembered as true.

Perhaps the Trump administration is emblematic of this problem, with some agencies and scientists seriously worried about the purging of information that the government collects. For example, suppose that the Trump administration wanted to fight climate change; they could simply modify the database to their heart's content.

You are probably already seeing how blockchain technology can factor into our 'post fact' society. There is a distrust brewing around the world in institutions; people are simply less likely to believe in authority figures and large groups that have for decades been an authority on issues ranging from the military to environmental concerns. This has been demonstrated most recently by Canadian scientists creating backups of databases detailing climate change.

The databases were originally collected by American agencies, and there is fear this data will be purged. Blockchain can solve the problem of a lack of trust in institutions, if only partially, by allowing the same information to be distributed around the world. There will not be fear of databases being hacked into, modified or changed because any such modifications would be stored in a blockchain. There would be an accurate record to keep databases up to date.

The widespread applications for shared databases are immense. From verifying the tax filing information for large corporations to checking the integrity of data collected by institutions, forging data becomes much more difficult with a blockchain foundation.

There is one use of blockchain that has already started to take off and is used by several organizations around the world, but the primary is in the hand so private citizens for now. Creating your own blockchain for practical use is quite easy with a plugin offered online (Multichain), and for right now the primary use in the hand of citizens is creating collated databases. For purposes of sharing large swaths of information that cannot be modified easily, this is by and far the best method.

It is worth mentioning that as blockchain is implemented in databases, it is likely to be a change that you do not even notice. It will merely be the underpinnings of how databases are managed. The hopeful outcome is that it will restore some faith in institutions, as people come to realize that data cannot be so easily faked.

❖ Digital Identity

Fraud is said to cost the so much and due to that, security is a top concern for businesses of all sizes. Blockchain technologies make tracking and managing digital identities secure and efficient, resulting in seamless sign-on and reduced fraud. Blockchain technology offers a solution to many digital identity issues, where identity can be uniquely authenticated in an irrefutable, immutable, and secure manner. Current methods use problematic password-based systems of shared secrets exchanged and stored on insecure systems. Blockchain-based authentication systems are based on irrefutable identity verification using digital signatures based on public key cryptography.

With blockchain identity authentication, the only check performed is whether or not the transaction was signed by the correct private key. It is inferred that whoever has access to the private key is the owner and the exact identity of the owner is deemed irrelevant. Blockchain technology can be applied to identify applications in areas like IDs, online account login, E-Residency, passports, and birth certificates.

❖ Online voting

The greatest barrier to getting electoral processes online, according to its detractors, is security. Using blockchain, a voter could check that her or his vote was successfully transmitted while remaining anonymous to the rest of the world. By casting votes as transactions, we can create a blockchain which keeps track of the tallies of the votes. This way, everyone can agree on the final count because they can count the votes themselves, and because of the blockchain audit trail, they can verify that no votes were changed or removed, and no illegitimate votes were added.

❖ Supply chain communications and proof-of-provenance

Most of the things we buy aren't made by a single entity, but by a chain of suppliers who sell their components (e.g., graphite for pencils) to a company that assembles and markets the final product. If any one of those components fails, however, the brand takes the brunt of the backlash and it holds the majority of the responsibility for its supply chain. But, what if a company could proactively provide digitally permanent, audible records that show stakeholders the state of the product at each value-added step? Blockchain aims to do just this.

The future potential of blockchain applications is still evolving. The next two to three years will all be about experimenting. Regardless of which application becomes the first on a global scale. The bottom line is, Blockchain is here to stay and is transforming how our society functions.

Chapter 4: Limitations of Blockchain

Before we can get into the limitations of this technology, it is important to consider a few benefits that it brings about. In case you're thinking about business opportunities for blockchain technology, it's useful to avoid getting over-excited about an idea that a traditional centralized database could do just as well. How do you parse which is which?

What specific benefits do blockchains bring compared to traditional databases?

Decentralized / shared control

This benefit comes in three forms. First, blockchains allow enemies to work together for a common benefit. It makes blockchains a political tool. Examples include: Open Music Initiative for the music labels. More generally, blockchains can reduce friction in forming consortiums, because the entities involved do not have to give control of the infrastructure to some single entity. So for some contexts, a blockchain is a "consortium database."

Second, when the network is decentralized and sufficiently open and for anyone to use, then it becomes a new public utility, like electricity, the Internet, or the Web. Examples include Bitcoin as an e-money or e-gold utility, Ethereum as a decentralized processing utility, IPFS as a decentralized file system utility, or IPDB as a decentralized database utility.

Third, interoperability protocols can further reduce friction, especially for public networks. If networks interoperate, then which network you're using matters less. You aren't necessarily stuck with your initial choice of network. The internet is a network of networks; do you care which sub-network you're in? Two key protocols are:

• *Interledger*. It connects two networks, even if those networks speak a different language. You can think of it like a well-defined protocol for an exchange. Or, like a router for the internet, but instead of handling just data, it also handles assets. How: escrow in the forward direction from source to target, then release funds backward. That's basically it. But it enables easy connection between Bitcoin, Ethereum, SWIFT, BigchainDB, and more.

• *IPLD (Interplanetary Linked Data).* This allows data blobs to flow through the walls of one network to another. It makes the networks permeable. How: an opinionated way to hash the structured data format JSON. That's basically it. But it means you can initially store your JSON data on, say, IPFS, but know that it can also flow to IPDB or whatever network supports IPLD.

Immutability/audit trail

When you write data to a blockchain, it's like etching the data into stone. This could be used for education credentials, land registries, and more.

If you have a series of transactions over time, you gain an immutable audit trail, which is useful for financial audits, art provenance, food history, and more.

Assets/exchanges

Once you have a data store that no single entity owns or controls, and no one can change what's already written, then it unlocks the possibility for assets themselves to live in the data store. Put another way: it gives benefits similar to double-entry bookkeeping but much easier to apply.

Once you have native assets, you've lowered the friction in implementing exchanges. Blockchain applications in traditional exchanges include stocks, currency, or energy. Multi-sided platforms are also exchanges; examples include social media, ride sharing, and online retail marketplace.

Why are people excited about blockchain?

BLOCKCHAIN

Almost everyone can agree that the blockchain is one of the most interesting and disruptive forces to come along in quite some time. And, that's because the blockchain is able to:

Prevents payment scams.

One of the most talked about advantages involving blockchain technology is how it can prevent future payment scams. For starters, it would protect both buyers and sellers by using "smart contracts." This procedure would avoid those instances where you purchase an item and the seller doesn't follow through.

Cuts out the middleman.

The blockchain is a peer-to-peer system, meaning that transactions are between you and another party. This simple two party only transaction could be a real game changer. We use this to be able to facilitate cheap ecash transactions across the world. For example, you could send friends or family money anywhere in the world without having to pay for the transaction or currency fees that traditional banking or financial institutions have used.

Settles transactions in minutes.

Imagine being able to send and receive money from across the globe in just a matter of minutes. How about receiving a signed contract or vehicle title in just a day? No matter the scenario, blockchain decentralized and the P2P system allows you to settle any digital wallet transaction quickly, as opposed to waiting days or weeks.

Increases storage.

Cloud storage is an incredible development. But, you don't have any control of the storage infrastructure. It's in the hands of Google, Dropbox, Facebook or Apple. And, that could become a concern if you value your privacy. Since you'll need an encryption key to access your data, you can rest assured that no one else can access it except you.

BLOCKCHAIN

Rewards users.

Who doesn't love reward programs? The blockchain can improve loyalty programs by giving customers the ability to trade points among each other since the transactions would be placed in the public ledger. It would also open up the possibility of using points at different vendors. For example, you could use some of your airline points at your favorite coffee shop or e-commerce site.

Because of those capabilities, the blockchain will be able to disrupt the following;

- Finance - Blockchain will remove the need for traditional banking and financial institutions by replacing back-office systems with a P2P system.

- Patents and Copyright - Whether it's a new innovation, gaming app or piece of music, the blockchain can prove that you had ownership of the intellectual property first.

- Voting - When people cast their ballots, it will be recorded during elections.

- Collectibles - The blockchain could be used to track and validate scarce or limited items like coupons or a piece of artwork.

- Bills of Lading - Cryptographic signatures can be used to eliminate distrust on everything from shipped products to changing shifts at work.

What are the limitations of the blockchain technology?

Emerging technologies can generate extravagant expectations. One of the latest developments promised to either save or destroy the world is the blockchain. The potential for the blockchain is enormous, but advocates, however, would be wise to understand the limits of the technology. Here are some limitations of the blockchain technology:

BLOCKCHAIN

Agreements among strangers

There is a major problem within the bitcoin network. It is a problem that manifests from the era in which Bitcoin was born, and is a problem that other cryptocurrencies have tried to address. The bitcoin network works based on a hashing system, where new information is generated with a string of data is then verified by the bitcoin network.

Right now the total length of the hash is predefined, and the length is suitable for generating the appropriate amount of data to be stored. The problem is that some point in the future bitcoin may have a crisis of differing public ledgers. As the hashing data gets more complex as a result of the increased complexity of the total number of transactions, there is a chance that the hash data will fork, meaning two separate ledgers will be created that do not match all of the past transactions.

While this problem is some number of years off from being a detriment to the currency, it brings into focus a large issue- the only way to fix this problem is to change the source code of the platform, but to do this you would need an agreement among the many people within the Bitcoin network.

It is from this singular crisis that we see the problem with having many anonymous users try and come to an agreement. To fix this problem is actually relatively simple and just requires a change in the protocol used for Bitcoin. There have been a number of proposals, and all of them would be successful to some degree.

At the very least they would push the problem further down the road. The reason why so many people cannot agree is simply because there are competing viewpoints about how to solve the hashing problem. Well known Bitcoin advocates have made suggestions that have merely fallen on deaf ears. The cryptocurrency network is one of the skeptics, and maybe even a few conspiracy theorists. There is a very real fear that changing the Bitcoin network would mean that one person has a distinct advantage over everyone else.

This is both a strength and weakness of the Bitcoin network. It is both rigid and flexible at the same time, and incredibly the downfall of the network might come down to politicizing the issue of changing a measure that must be changed for the survival of the network.

Cryptocurrencies currently aren't practical

The growth of the Silk Road and other dark websites to buy illicit substances gave rise to a number of companies that handle the exchange of cryptocurrencies into real, government-backed currency. This was a key dilemma for bitcoin, and it continues to be one of the main limitation of the currency.

The market of exchange determines the value of a single bitcoin, but right now there is no central market for exchange. Unlike Forex networks of a government exchange program, it is much more difficult to get an accurate idea of how much a Bitcoin is worth. In the early days to exchange Bitcoin for the government-backed currency, you needed to find someone that would want to trade dollars for Bitcoin.

Obviously this is a horribly inefficient way to make a currency conversion, but even more importantly how would you know what a bitcoin is even worth? There needed to be exchanges where this information was kept, and without any central institutions furthering the interests of Bitcoin, this was left up to private enterprise.

From the very beginning currency exchange with cryptocurrencies has been a nightmare. The companies that have done a "terrible" job of transparency, accountability and even offering stable services throughout the years have been shut down for fraud; others have been shut down for their dealings with the Silk Road.

Ironically a distributed ledger is perhaps the greatest way of sharing information, but Bitcoin and other cryptocurrencies face an existential issue of an information problem concerning the value of the currency.

As of today, there are limited places where you can use a cryptocurrency outright. You will likely need to exchange a cryptocurrency for a government-backed currency, but this is both difficult and a poor financial decision. Bitcoins have very limited practical use and so their value fluctuates greatly because the main driver of perceived value is the speculation that Bitcoins will be worth something in the future.

The more popular the cryptocurrency, the larger an issue this becomes. Dealing with exchanges for currency is difficult because so few can be trusted, particularly if you are asking them to hold onto you Bitcoin wallet. One case comes to mind where a Japanese exchange founder attempted to cover a hack where more than a half of the exchange's total Bitcoins were stolen.

This was eventually litigated and the exchange was shut down. The point is that the safest place for your Bitcoin or other cryptocurrency is merely to keep it on your persons. Additionally, the most profitable thing you can do with a cryptocurrency is to merely hold onto it. Right now the uses are merely theoretical, and so the practicality of any cryptocurrency is extremely limited.

Smart contracts require more technologies

In the previous chapter, we learnt about the cost saving technology that will come in the form of smart contracts. This technology will first be seen in website-to-website interactions, such as ad revenue and settling disputes where transactions were handled digitally. For this technology to really take off, it requires the use of geo-tagging physical assets.

This technology is coming, but there will be a huge hurdle in widespread adoption. To make effective smart contracts, geo-tagging has to be done on all items pertaining to a contract. In the meantime, smart contracts will be relegated to only online dealings.

Government cryptocurrencies

There is a very real possibility that in the future the government will adopt the foundational technology behind Bitcoin to roll out their own digital currency. There have been a number of theoretical models for this change, ranging from keeping a public ledger to simply putting a single ledger in the hands of the government.

This is not so much a limitation of the currency, but rather an eventual outcome and could possibly be the downfall of independent cryptocurrencies. Right now the value of cryptocurrencies deals with their rarity, but also the value that traders have put in the underlying technology. Once a government issues its own currency that uses the same base technology, it has been predicted that the value of all other cryptocurrencies will fall precipitously.

The crisis would appear if multiple governments switched to a blockchain method of issuing currency, and made it impossible to trade independent cryptocurrencies for government-backed ones. Again, this is not the most serious concern, and the market for cryptocurrencies will be fine in the future.

While the values may change, their nature of being similar to gold will mean that their rarity will also leave individual currencies with some value, if they do depreciate over time.

Energy

The blockchain has been heralded as a possible solution to global warming, by providing a transparent currency that isn't based on consumption. But the distributed ledger technology leaves a massive carbon footprint of its own. The computing power required to processes bitcoin dwarfs that of the world's fastest 500 supercomputers combined.

Interoperability

One of the concerns around building standards for that network of data is how to make sure it's interoperable, and not just a bunch of stuff. The fragmented landscape of competing blockchains has failed to produce international standards for the technology. Greater interoperability is needed to make the blockchain compatible with the wider web and to integrate them into existing practices and processes.

Privacy

Data on the blockchain is inherently shared publicly between everybody on the system. That level of openness isn't always a sufficiently secure approach to data storage. The Bitcoin database has every transaction that has been carried out in that database, and everybody has a copy. Who would publish their bank statements online openly for everyone to see? There have already been some unnerving applications of the blockchain.

Changing truths

The indelible nature of information in the blockchain implies that truth is eternal. The reality is often grayer than such absolutes. The right to be forgotten is enshrined in EU law, but difficult to apply to the immutable data store of a public blockchain. There are various bits of legislation in the UK that say if you change your gender you have the right to retroactively apply that all the way back through history. If your gender is stored inside a public blockchain of driving licenses or land registry, how do you do that?

Illegal information

If illegal data is embedded into the blockchain that could hypothetically make the entire blockchain illegal and everyone on it guilty of breaking the law. Now it's on the blockchain and is on everyone's machine. Nobody cares anymore, but the point remains.

Encryption

Encryption of blockchain data creates a number of issues. Anyone with the encryption key can read the encrypted data if the key is made public, while if the key to unlock the blockchain is lost you can never get it back.

BLOCKCHAIN

Any encryption used will be broken eventually, whether through loopholes, backdoors, or new technologies. If you're looking at applications of blockchains It's really important to note that they do not solve privacy issues, they're only useful for confirming authenticity.

Finding information

Adding information to the blockchain is only the beginning. To use the data, there needs to be a reliable way to find it. The blockchain can be indexed into searchable databases to, but to function reliably, every user will need to have that entire blockchain stored and also require a larger search index to be built from it. You can have one site that has a search index for the chain that would be fine but then trusting that one site is an issue. So how do we have distributed search indexes? Things like that start to become issues as well.

Chapter 5: Profiting from Blockchain

Profitability in mining and trading

The long-term growth of cryptocurrencies has shown that it is a valid avenue for investment. If someone invested just a few thousand dollars in Bitcoin in 2010, they today would have made several million dollars. While it is true that the values of currencies are highly volatile, provided you have the ability to wait for the ideal moment to sell a currency, it is still quite possible to make a great amount of profit through selling cryptocurrencies.

The question then becomes how can one gain access to cryptocurrencies, and what is the cheapest way to acquire them? Right now there are two avenues for acquiring a cryptocurrency; one can either mine them or buy them from someone that already has some. Perhaps if you are trying to make profit through these currencies, you try and mine some yourself. It is quite an easy process to set up a number of mining "rigs".

The software that will mine for currency is also easy to set up, and over years of mining develop ways of cutting down on overhead costs and ensuring that each dollar spent is used efficiently as possible.

The other avenue is to acquire cryptocurrency is to buy some from someone that has mined their own, it bought it themselves. This certainly can be profitable, the long-term profitability in this method is a little bit more difficult, as the fluctuation in prices is very severe and the main method of making money turns into something of a mimicking a commodities trader.

You must time when to buy and when to sell a currency, and be sentient of the market conditions when making a transaction. Additionally, with the current cost of mining the popular cryptocurrencies, it is simply cheaper to mine it yourself than to buy any significant sum. For example, you could easily spend two to three thousand dollars on cryptocurrencies- an investment that will pay off in a few years. If you spent the same investment on mining rigs, the profit potential is unlimited.

You are not buying into a specific currency, but rather than betting on the hardware that will mine any number of cryptocurrencies. If the market shifts and one currency becomes more favorable than another, you can simply switch the primary currency that you mine. As of today, you can mine NXT and Bitcoin primarily; these two have shown to be the most profitable, however, there are a number of options available to miners.

Deciding on a currency to mine

There is an argument that the best currencies to mine are the ones that are the easiest to obtain. For example, depending on the length of time that a currency has been around, coupled with the underlying algorithm for generating new blocks in a blockchain, some currencies can be mined much faster than others. What this argument fails to take into account is that a currency must be popular for it to sell at a good price.

You might be able to mine thousands of coins from a lesser known currency, but unless that currency picks up in value, then it will have been a wasted investment. You need to be able to sell a currency to make any profit. For this reason, there is a list of the best currencies to mine below. This is specifically for mining currencies, it is not recommendable to buy any of the currencies wholesale with the money that you already have.

The following currencies offer a mix of in-demand that are relatively easy to mine, and have decent exchanges so you can cash out at your leisure.

1. Bitcoin
2. Ripple
3. Litecoin
4. Bitshares
5. Darkcoin
6. Nxt
7. Dogecoin
8. MaidSafeCoin

BLOCKCHAIN

9. Stellar
10. Paycoin

Some of the Best Bitcoin Business Ideas & Opportunities

If you are an entrepreneur or would like to be one then profitable Bitcoin business ideas are not hard to come by. That doesn't mean that it will be easy for you to make money, or that you are guaranteed success. But that first step of finding interesting opportunities that are worthy of consideration should not be a barrier to anybody willing to invest their time and/or money into digital currency. Here are some of the ways to profit from the blockchain technology:

Becoming a Bitcoin Broker

Perhaps one of the most obvious as well as one of the most popular ways to start a business in this industry is to set yourself up as a broker, buying and selling coins to other users. Unlike other areas of finance, digital currency users often have a preference for using peer-to-peer services rather than large companies.

This preference extends to exchanges, meaning that it is very easy for a small trader to set themselves up as a broker in their local area or over the internet. As a broker, you earn your profit from the 'spread' – the difference between bids and ask prices. This varies according to market conditions and the payment method you are using, but you can get a rough idea simply by visiting the buy and sell pages on the site for your local area.

Retail Businesses

You can buy most things with Bitcoin today, but there are still opportunities available for new retail businesses which accept digital currency payments to make a name for themselves. The low transaction costs and freedom from chargebacks make BTC payments an attractive proposition for retailers, and if you can pass on some of those savings to your customers in the form of discounts you have a great chance to attract new business.

Perhaps the easiest way to set up a new retail business and take payment in BTC is using an internet shop builder service like Shopify.

Bitcoin Mining

Mining is a very competitive business today, but that doesn't mean that it is impossible for new start-ups to succeed. The key to being able to make a profit from mining is that you need access to low-cost electricity. Setting up in an area with a cold climate may also help to reduce equipment cooling costs.

This is, however, a risky business that is dependent on factors which are difficult to forecast (such as the price of BTC) and which will probably require a high capital outlay for equipment, so make sure you really know what you are getting into before setting up shop as a miner.

Consultancy Businesses

You probably know more about bitcoin than 99% of other people, at the very least. So why not put that knowledge to good use by helping other businesses? Both Bitcoin itself and the blockchain technology which underpins it offers a wealth of opportunities, not only for setting up a new business but also within established businesses. Unfortunately, most companies just don't know how to take advantage of them. This doesn't just stop at accepting Bitcoin payments, it could involve using the blockchain for low-cost notary services, as an asset registry, smart contracts and a lot more besides.

BTMs: Operating a Bitcoin ATM

If you have enough capital behind you then a more easily scalable, and potentially more profitable way to set up a business buying and selling coins, is by operating specialist 'Automatic Teller Machines' (ATMs) sometimes known as 'Bitcoin Teller Machines' (BTM).

Fees charged by BTMs seem to start around the range of 5-10% per transaction, and in some cases are a lot higher. Operators who manage to get their machines into the best locations often report ROI for their initial capital in less than a year. These machines do not take up a lot of space, so renting locations doesn't need to cost the earth. But with the cost of the machine itself, and the requirement to stock it with notes, the initial outlay can be quite high.

White Label Casinos

A white label business is when another company allows you to take their product or service, rebrand it under your own name, and present it to the public as an independent business. Although the core product is not unique, these services often allow for a relatively high degree of customization.

Gambling has always been one of those areas in which the advantages of digital currency are most apparent. One of the reasons for this is because many countries do not classify it as being real money, which means that strict laws and regulations controlling online gambling may not apply to casinos which use BTC exclusively.

If you fancy running your own casino, poker or betting site then there are many white label opportunities for you to take advantage of. These can range from a complete 'turnkey' website which just requires you to add your own branding (and make sure that you are complying with local laws) to individual games that you can add to your own site.

Flipping Websites, Apps & Businesses

Business flipping is when you buy a business, increase its profitability, and then sell it on in a relatively short period of time. The term is more commonly used for online businesses in the form of website or app flipping, but can also be applied to bricks & mortar businesses.

There are many different websites and apps which could benefit from integrating digital currency into what they offer. It is also possible to buy websites and apps for much less than most other businesses, and to 'flip' them within a fairly short period of time. Integrating digital currency for in-app purchases or for user-rewards schemes, or simply converting stores to accept BTC payments, may be worth considering as ways to add value to an established business.

BLOCKCHAIN

Chapter 6: Building a mining rig

Mining basics

Thousands of computers all over the planet mine cryptocurrencies and many of these computers are doing so because of viruses designed specifically for mining. If you have ever had your computer slow down because of a virus, it is possible that a miner was installed on your computer.

The reason that someone would design a virus to do distributed mining is the cost of the hardware and the cost to power the hardware. A virus bypasses both of these costs so that the creator of the virus can only profit. Even so, it is unlikely that the creator of a virus that mines cryptocurrencies have made a tremendous sum of money.

You need very specific hardware to mine cryptocurrencies at an efficient rate. In this chapter, you will learn about all of the parts that you will need to build your own mining rig. It might seem like a monumental task, but the truth is that getting a mining rig off the ground is a relatively simple process.

All you need to do is select the parts and build it yourself, or barring that ask for help from a local computer store to help you assemble the parts- this was the avenue when you take your first mining rig, and along with paying for assembly you pay to learn how to build future rigs by yourself.

The key to successful cryptocurrency mining is to buy parts that are power efficient and capable of mining at a fast rate. Since these computers are designed for a singular purpose, there are many standard parts that you will not need. They are also based on desktop parts as these are cheaper, but they also allow for larger graphics processing units (GPU) - the basis for mining efficiently.

The reason that a virus will not be able to mine efficiently, even when on thousands of computers, is that most personal computers today are laptops. Laptops have a central processing unit (CPU) that handles the standard calculations, but also renders all the video and images that you see on your computer monitor.

The fact of the matter is that running calculations on a CPU is incredibly inefficient. A moderately powered graphics card will be able to mine at a far greater rate than even ten or twenty decently powered laptops.

It is important that as you build your mining rig that you spend money on the parts that are important, and minimize the costs on superfluous items. For example, you will not need computer monitors, keyboards and mice or speakers for these computers.

Ideally, they do not have more than the parts necessary for efficient mining. You will, of course, need one monitor and a single keyboard and mouse to set up the mining on machines, but once the setup is complete the rest is handled automatically.

Motherboard

Assembling a modern computer might seem terrifying, but the truth is that it is quite similar to assembling Lego. Each part is a block that plugs into a central board called the 'motherboard'. This base component is where you will put your CPU/GPU and all of the other components that make up your computer. A motherboard is a necessary part, but not one that you should be spending all that much money on.

Even relatively inexpensive motherboards will serve fine for the purposes of mining. The most important aspect is that you buy a motherboard that is full size, as opposed to micro. The main consideration here is heat, and a smaller board does not fare as well as a larger one.

You can expect a motherboard to cost between $50 and $80, with preferable brands being ASUS, GIGABYTE, and MSI. What is important is that your motherboard will be a determining factor in the type of central processing unit that you will use.

There are two main suppliers of processors, AMD and Intel. Depending on the motherboard, you will need the corresponding slot type and manufacturer. For example, an Intel LGA1155 board will only take Intel processors of that socket type. Do not let 'LGA1155' scare you- simply put marketers have not realized the usefulness of catchy names for computer parts.

It is advisable to go with an Intel based motherboard capable of fitting an i3 or i5 processor. AMD processors are simply worse at handling heat, which is one of your main concerns for reducing the overhead cost of running a mining rig. You may first want to look at your choices of CPU and then base your motherboard decision off of what processor you will be purchasing.

As a general rule of thumb, processor costs are more fixed than motherboards. You may find an excellent sale on a motherboard, but it is unlikely that the cost of a processor will change. Typically the manufacturer of CPUs will issue parts and never adjust the prices, only modifying pricing when new models of processors come out.

You should note that a motherboard will have many components already built into it. For example, your networking capability should be built into your motherboard. This is where you will also find your audio outputs and inputs for keyboard and mouse.

Power supply

The power supply is the most important component of a mining rig. There are countless stories you might have heard about people purchasing cheap power supply only to see their mining rig burst into flames. This is not meant to be an alarmist- power supplies are very safe, but they range in build quality greater than any other part on the list. The power supply is so essential because the computer will be running twenty-four hours a day, so stability is key.

The recommendable power supply includes EVGA, COSAIR, Or Thermaltake. They come in different wattages but for your needs, you must simply look for one that is 400 watts or greater. You should also note that power supplies are graded in terms of quality. Provided you buy a power supply that is silver or gold rated, you will be fine.

Additionally, the manufacturers listed above only make high-quality power supplies, so even a bronze rated EVGA power supply will be sufficient. Expect the cost of a power supply to range between $50 and $80, depending on sales and promotions.

Power supplies should be changed out every eighteen months. All power supplies fail; the question is merely a matter of time, and how they will die. A poorly manufactured power supply will take all of the connected parts with it, meaning that an entire rig can be lost. A well-made power supply will simply stop working, meaning you can salvage all of the parts I the machine. You would simply need to replace the power supply to get it running again.

Since these machines are running at maximum load all of the time, an eighteen month period of use is to be expected. You can run them for longer, but you always risk a failure that may take out other parts with it.

Graphics Card (GPU)

The graphics card is the second most important part of your mining rig. This is where the true computational power comes into play. While a central processing unit can only handle a few calculations at a time, a graphics card can handle hundreds at the same time. It is a byproduct of their design, taking instructions from the CPU to power onscreen images. The relative power in each core of the GPU is low, but with so many of them, they can mine much more efficiently than any CPU.

There are two main brands to buy from for graphics processors, NVidia and AMD. This is the same AMD that also makes COUs, and for what it is worth the other main CPU manufacturer, Intel, owns NVidia. Both companies have good models of consumer video cards for mining. Most recently both companies have released new products that consume less power and are more efficient when it comes to heat.

There is a bit of a trade off in these two manufacturers. NVidia costs a bit more, and is not as fast as similar AMD cards, however they are more reliable overall, consume less power and produce drastically less heat. NVidia cards were easily available before the latest line of AMD cards.

It is recommendable either to use a NVidia 1060 ($250) or an AMD 480 ($200). Both of these models have variations depending on the manufacturer of the card. For example, you may see an EVGA 480, but this is the same base product as the one that AMD offers.

Without getting into too many details, the boards are free to be manufactured provided the companies get royalties for the technology. The primary difference is sometimes the clock speeds are increased, or the mounted cooling units are more efficient. There is no brand of either GPU that is not recommended so you can go with whatever is cheapest.

In these models, you will also see that they are differentiated based on the amount of memory that they have, coming in 3 and 6 GB versions for NVidia, and 4 and 8GB versions for AMD. For the purposes of mining, the onboard video memory does not matter- buy the card with less memory as it will always be cheaper.

A GPU should be replaced once every two to three years, but the main catalyst for replacement should be that a new and more efficient model has been released. When a GPU dies, it simply stops working, so the build quality is not as essential as with the power supply.

It is recommended that if you start building mining rigs, you on occasion look at benchmarks for power efficiency in video cards. New cards come out roughly every year or so, with major revisions happening every two to four years- the latest major revisions were in 2016.

Central processing unit (CPU)

For processors, you have a choice between Intel and AMD. What processor type you buy does not matter that much, as both are equally suitable for mining. What is important is that the socket type of the processor matches the motherboard that you buy. With this, it is advisable to buy your motherboard and CPU at the same time, if only to ensure compatibility.

Like with video cards, the differences between the types of CPUs come down to heat and cost. AMD costs less but produces more heat. Intel costs more but produces less heat. Remember that more heat means more power consumption, and so you can trade the upfront cost for long-term savings by going with Intel, or you can pay less now and go with AMD.

This is one of the more expensive parts of a mining rig, with the average cost of a processor being slightly over $ 200. The naming convention of the AMD processors is not an intuitive as the Intel system. An i3 processor is recommendable which could cost as low as $ 150, or any AMD processor that is 'Bulldozer' or newer.

The upper limit for the cost of this part is $250, with the cheapest processor type coming in at $ 150. Other guides may suggest going with an i5 Intel processor, but the truth is that you don't really need this type of power. We are building a mining rig for home use, not a rig that will be joined with hundreds of other computers in a warehouse.

Memory (RAM)

The random access memory, or RAM, is a part that should not cost you more than $40 or $50. There are many different manufacturers of RAM, and one cannot one brand over the other because there is a consistent quality to ram, which has more or less made it a commodity. You will need between 4 and 8 GB of system ram for efficient mining. 4 GB is fine in 2017, but by 2018 or 2019 you will likely want to have 8 GB of RAM in your system.

This is an easily upgradeable part, so one can buy the cheapest and smallest amount as they get started. Please note that while the numbers and markings are the same, system RAM is very different from video card memory. One does not substitute the other, so even if you have a 3 GB video card, you will still need a 4GB of system RAM for your rig.

Storage

Storage is an essential part of any computer, but no one that you should spend a lot of money on storage options come in the form of hard drives (HDD) and solid state drives (SSD), with later being far more expensive. There has been a push for miners to move to SSD due to the lower power usage, but the cost of SSDs is high that the upfront costs do not justify the long-term power usage.

Any hard drive from any manufacturer will do – look for storage sizes of 80 GB up to 1TB, and buy whatever part is cheapest. Note that hard drives do fail and that without a storage drive your rig will stop working. You will need to replace it if it fails, but you will not lose any of your cryptocurrency, as that is stored on the public ledger and not locally on your machine.

Operating system

Your options for operating systems are Windows or Linux, and Linux is highly recommended. A Linux license is free and an image can be found online quite easily. To install an operating system like Linux, simply create a bootable flash drive by following the instructions provided by the specific version of Linux that you install. Linux Red Hat or Ubuntu are highly recommended- you can think of these naming conventions as Windows 7, 10 etc.

They all have the same foundation but their interface is slightly different. Before you decide on the version of Linux that you want, you should look at the compatibility chart to ensure your cryptocurrency is listed. Once you have Linux installed, you must simply download the cryptocurrency client that you wish to mine. Install the client, set up an account and follow the instructions to start mining.

Conclusion

This book has tried to demonstrate that blockchain technology's many concepts and features might be broadly extensible to a wide variety of situations. These features apply not just to the immediate context of currency and payments, or to contracts, property, and all financial markets transactions, but beyond to segments as diverse as government, health, science, literacy, publishing, economic development, art, and culture, and possibly even more broadly to enable orders-of-magnitude larger-scale human progress.

Blockchain technology could be quite complementary in a possibility space for the future world that includes both centralized and decentralized models. Like any new technology, the blockchain is an idea that initially disrupts, and over time it could promote the development of a larger ecosystem that includes both the old way and the new innovation. Some historical examples are that the advent of the radio, in fact, led to increased record sales, and e-readers such as the Kindle have increased book sales. Now, we obtain news from the blogs, Twitter, and personalized drone feeds alike. We consume media from both large entertainment companies and YouTube. Thus, over time, blockchain technology could exist in a larger ecosystem with both centralized and decentralized models.

There is little doubt that blockchain technology will greatly change the economy in the coming years. From revamping the way that information is shared across organizations, to possibly being the underlying technology for a government issued cryptocurrency, the strength of blockchain lies in its ability to force cooperation among many disparate parties.

As Bitcoin and the blockchain continue to progress, it is noteworthy to keep a close watch on the blockchain, given its capabilities to innovate various areas that are relevant to the society as well as the day-to-day lives in the digital age. The progression of Bitcoin and the blockchain would revamp and enhance the following concepts:

- Globalism - Bitcoin and the blockchain address globalism with mobile applications and exchange platforms that facilitate financial and economic activities worldwide.

- Security - The blockchain technology addresses security in areas including insurance, law, and data security by validation of the information in the blockchain ledger.

- Democracy - Bitcoin and the blockchain address democracy by reshaping the functions of governments, organizations, and corporations with commercial influence in addition to the Blockchain's ability to make voting systems more effective.

Blockchain technology can really be applied to not just a cryptocurrency like bitcoin, but to any "asset" that can be stored, distributed or transacted – property titles, music, insurance, physical goods and assets, even your data. This technology has great implications for the financial services industry as well. On implementing a decentralized database or a public registry like blockchain to verify the identities of all parties, no longer will we need to have our transactions stay "pending" for three days. The settlement would be instantaneous since the transaction and settlement would happen simultaneously once the ledger is updated. There are many such use cases.

You are now well aware of where the future is heading and what you can expect from the future of blockchain technology. You also have a firm understanding of how cryptocurrencies work, and how you can make a profit by mining them yourself.

The material in the book is compiled for the reader to educate him or herself on blockchain and to make an impact on that reader's life. You should now feel enlightened about the nature of blockchain, and that you find some of the same amazement that others have over the years, about the genius of its creation and the interesting nature of its creators. It is truly an invention of the Internet era; a technology to restore power to the people.

By spreading information without a central institution or authority, blockchain democratizes information in much the same way the internet originally did. It serves as a platform for services, ranging from databases to currencies, but its uses will grow far greater than anything we have seen in its current form today.

The last chapter of this book is dedicated to the idea of mining cryptocurrencies for profit. For some readers, this might seem fanciful, as if it is reserved for only the most technically inclined. You can be assured that the only thing necessary to mine cryptocurrencies is the motivation and desire for profit. Most people came to blockchain technology purely out of fascination, but after learning and understanding how blockchain works, they have found that mining is a simple process that can be used to supplement your income.

It is advisable to review chapter six and consider building a mining rig; the cost of investment is low, and the potential for profit is unlimited. A mining rig is a platform to generate profit for many years into the future. While it is true that the exact amount of profit is unknown, cryptocurrencies have proven to be overall resilient enough that they are worth the investment.

www.ingramcontent.com/pod-product-compliance
Lightning Source LLC
Chambersburg PA
CBHW050024230526
45470CB00003B/1118